The Heritage Collection

REATHA CLARK KING
Scientist, Educator, Changemaker

Rosemond Sarpong Owens

Reatha Clark King : Scientist, Educator, and Changemaker

Copyright © 2025 by Rosemond Sarpong Owens

Layout designer: Nassim Sarkar

Illustrator: Amina Yaqoob

Library of Congress Control Number: 2025903186

All rights reserved.

No part of this publication may be reproduced, stored in a retrieval system, a database, and/or published in any form or by any means, electronic, mechanical, photocopying, recording or otherwise, without the prior written permission of the publisher.

ISBN 978-1-956051-29-2 (hardcover)
ISBN 978-1956051-30-8 (ebook)

Published by Lion's Historian Press
https://www.lionshistorian.net/

To all young people — May you embrace every chance to learn, grow, and chase your dreams. Stay curious, stay eager, and never stop believing in the power of knowledge. The future is yours to shape.

"What I would hope for, and I will be wishing for this every day, is that the young people today will seize the opportunities to learn, and be eager about learning."

—Dr. Reatha Clark King

EDITORIAL REVIEW

Owens has beautifully captured the life and legacy of Dr. Reatha Clark King. This book is an incredible gift to younger readers, who may not realize the profound impact Dr. King has made in science, academia, business, and philanthropy. Dr. King's story of beating the odds and blazing new trails represents an American ideal that has the power to inspire young readers to strive, to speak up, and to step boldly forward as they carve their own paths in a world that isn't always fair or just.

Kate Leibfried- Author, "Find a Trail or Blaze One: A Biography of Dr. Reatha Clark King"

CONTENTS

Chapter 1: The Beginning Years 1
Chapter 2: Curious Reatha 3
Chapter 3: Faith & Family 5
Chapter 4: School 7
Chapter 5: The Scientist 9
Chapter 6: Breaking Barriers 11
Chapter 7: Where The Heart Is 13
Chapter 8: A Giant Leap 15
Chapter 9: A New Path 17
Chapter 10: The University President 19
Chapter 11: The General Mills Foundation 21
Chapter 12: Still Going Strong 23
Chapter 13: A Lasting Legacy 25

Photo Gallery 27
Important Dates in the Life of Dr. Reatha Clark King 34
Awards and Recognition 35
References 36
Glossary 37
Acknowledgements 40
About the Author 41
Other Books in the Heritage Collection 42

The Beginning Years

Reatha was born in 1938 during the Great Depression. The world was very different and there were no cell phones, no computers, and no astronauts in space. In the small town of Moultrie, Georgia, Reatha, the daughter of a sharecropper, spent hot summer days working in the cotton fields with her family.

Picking cotton was difficult, but Reatha remembered what her grandmother Mamie always said: "Whatever you do, do it well." She set a big goal to pick 200 pounds of cotton every day and she accomplished her goal because her nature was to do everything well.

> "I remember my childhood as one that was built on perseverance. We were poor, and there was no time for play. My family worked in the cotton fields—something that built character and instilled in me a strong work ethic."

Curious Reatha

Reatha loved being outdoors. She was fascinated by how birds made their nests and how the bugs moved around. She would always ask questions like, "Why is the sky blue?" or "How do plants grow tall?" She dreamed of understanding the world more every day. Reatha had a natural curiosity and love for solving problems which would play a huge role in shaping her future career path.

> "From the moment I could ask 'why,' I wanted to understand the world around me. Curiosity was my first teacher, and it led me to places I never imagined—from the cotton fields of Georgia to the laboratories of NASA."

Faith & Family

Reatha adored her sisters Mamie and Dot. They grew up in a large, loving family. Church-going was an essential part of their family, and Reatha and her sisters loved dressing up and attending church on Sundays. The church was not a just place for singing and praying. It was also a meeting place for the community to connect. With parents who cheered her on, a strong church family, and a neighborhood that cared for her, Reatha and her sisters flourished.

> "Faith is what sustained me. When you're faced with seemingly insurmountable odds, it is your belief that you can rise above that keeps you going."

School

Reatha loved school and attended an all-Black school because of segregation. She studied hard and her teachers encouraged her to explore new ideas. One day during Negro History Week, Reatha learned about George Washington Carver, a brilliant scientist who made amazing discoveries. He found over 300 uses for peanuts, including making peanut oil and soap, and helped Southern farmers improve their soil so they could grow better crops. His story inspired her.

A tiny seed of possibility about becoming a scientist began to grow. With her love of learning and determination, Reatha's efforts paid off. She graduated as a valedictorian and received a scholarship to Clark College in Atlanta, a big step toward making her dreams come true.

> "My inspiration comes from the sisters and teachers who had such great influence on my life."

Note for Readers:

In the past, Negro History Week was celebrated every February to honor the achievements of Black people. It was later expanded and became Black History Month, which is now celebrated throughout February. The term "Negro" was commonly used at that time, but today, we use the term Black or African American to refer to people of African descent. It's important to understand history and language as they change over time.

The Scientist

Reatha never thought she would attend college. She was excited when she arrived at Clark College in Atlanta and was amazed by her surroundings.

At Clark, Reatha learned a lot, not only from her studies but also about herself. Reatha fell in love with chemistry. She became a research scientist and a chemist, proving to herself and others that anything was possible. She broke the norm. At that time, women were told that the appropriate work for them was to be a nurse, social worker, or teacher, particularly a home economics teacher.

> "In a field dominated by men, especially white men, my perseverance was often tested. But every challenge was an opportunity for growth."

Breaking Barriers

Reatha's love for science led her to the University of Chicago, where she worked hard in the lab and made exciting discoveries. In 1960, she became the first Black woman to earn a master's degree in Chemistry from the school.

But she didn't stop there! In 1963, she earned her PhD in Chemistry, officially becoming Dr. Reatha Clark King. At a time when few Black women were in science, she proved that anything was possible.

"Being a scientist prepares you for life and leads you to do work of high quality."

Chapter 7

Where The Heart Is

Reatha married her college sweetheart, Judge King. Her husband became her rock. He was very supportive, and he believed in her dreams, helping her face challenges with strength. Together, they had two sons, Jay and Scott. Just like Reatha's large, loving family she grew up in, she and Judge created a home filled with love.

"Through every challenge, love and family remained my guiding light, shaping my journey as a scientist, educator, and leader."

A Giant Leap

At the age of 25 years, Reatha Clark King made history as the first African-American female chemist at the National Bureau of Standards. She was instrumental in devising the fuel system of the Apollo 11 mission, making the moon landing possible. Reatha had invented a special tube that stopped the rocket fuel from exploding. This was important for the success of the mission. When she heard, "One small step for man, one giant leap for mankind," Reatha felt proud and touched, knowing her work contributed to this important moment.

> "It was a thrill to know that the work I was doing contributed, even in a small way, to putting men on the moon. I often reflect on how science can open doors, not just for individuals, but for humanity."

A New Path

After five years as a chemist, Reatha made a career change into education. She wanted to help students reach their goals. She started teaching chemistry and became an associate professor at York College. After only a year, she was named Associate Dean for Natural Sciences and Mathematics. She was the first woman of color to hold that job at the school. Reatha moved up the ladder to become the Associate Dean of Academic Affairs. Reatha continued her own education and even earned an MBA degree.

"Education is the key that unlocks opportunities. Without it, you can have no future. My work in education has always been about making sure that others can find that key."

The University President

Reatha broke barriers when she became the first African-American woman to become president of Metropolitan State University in Minnesota.

In her new role, Reatha worked hard to make the university a place where students from all backgrounds could succeed, bringing together innovative ideas to create a school with a clear, strategic path. Reatha's leadership helped to make higher education in Minnesota a place of opportunity for everyone.

> "I'm leading toward a cause: to get more opportunities for people. It is in my blood to remove unjust barriers and help people appreciate themselves and be who they are."

The General Mills Foundation

Dr. Reatha Clark King loved helping others. She believed that strong communities made the world a better place. That's why she joined the General Mills Foundation, the part of General Mills that helps people by supporting schools, health programs, and the arts. General Mills is known for making Cheerios, Pillsbury biscuits, and Betty Crocker cakes, but they also give back to communities in big ways!

One of Reatha's biggest contributions was co-founding the Twin Cities Martin Luther King Jr. Breakfast in 1991. Every year, thousands of people come together at this event to celebrate Dr. King's message of kindness, equality, and service.

Reatha knew that helping others was one of the greatest things a person could do. She used her leadership to make a difference, inspiring people to give, grow, and help their communities thrive just like Dr. King.

> "I feel blessed. However, it compels me to want to give back. Because I feel indebted to those who helped me. I need to be a champion now, to open doors for others."

Still Going Strong

You would think that after such an accomplished career, Reatha would slow down, but she continues to stay busy. She spends her time with her family and her grandkids, who bring her so much joy. She is also active in her church and continues to serve on various boards, using her experience to support her community. Plus, she still takes the time to speak to young people, inspiring them to dream big and work hard. Even in retirement, Reatha remains as engaged and inspiring as ever.

> "True leadership is about service. It's about bringing others along with you and ensuring that those who come behind you have the same opportunities to succeed."

A Lasting Legacy

Reatha Clark King has not only scaled mountains, she has conquered them. From the steep cliffs of illiteracy and poverty in her youth to the towering peaks of science, education, and leadership, Reatha's journey has been one of unyielding courage.

With each challenge, she lifted herself up, and with every step forward, she pulled others along. Whether it was breaking barriers as a research chemist, carving new paths in the world of academia, or becoming the first African-American woman president of Metropolitan State University, Reatha's life is proof that no mountain is too high to climb.

Her grandmother Mamie's words, *"You lift as you climb,"* were not just a guiding principle; they were a way of life for Reatha. She believed that success wasn't just about reaching the top but about helping others rise alongside you.

And while Reatha has already climbed so many mountains, her story is far from over. Her journey continues to inspire, showing that every climb, no matter how steep, is a chance to reach new heights and bring others along for the ride.

The mountain Reatha climbed is now a path for all of us to follow. What mountains will you climb, and how will you lift others as you go?

The End

"My legacy will not just be in the work I did, but in how it helped others, especially young Black students realize their potential."

PHOTO GALLERY

Credit: Scott Clark King

PHOTO GALLERY

Credit: Scott Clark King

PHOTO GALLERY

Credit: Scott Clark King

PHOTO GALLERY

Credit: Scott Clark King

PHOTO GALLERY

Credit: Scott Clark King

31

PHOTO GALLERY

Credit: Scott Clark King

PHOTO GALLERY

Credit: Rosemond Sarpong Owens

IMPORTANT DATES IN THE LIFE OF DR. REATHA CLARK KING

- **April 11, 1938** — Born in Pavo, Georgia, to sharecropper parents.

- **1960** — Earned a Master's degree in Chemistry from the University of Chicago.

- **1963** — Became the first Black woman to earn a PhD in Chemistry from the University of Chicago.

- **1963-1968** — Worked at the National Bureau of Standards, where she contributed to the U.S. space program by developing a coiled tube to cool rocket fuel.

- **1968-1977** — Served as a professor and later Associate Dean at York College, City University of New York.

- **1977-1988** — Appointed as the second president of Metropolitan State University in Minneapolis, significantly expanding the institution's programs and student enrollment.

- **1988-2002** — Served as Vice President of General Mills and President & Executive Director of the General Mills Foundation, leading philanthropic efforts.

- **2002-Present** — Continues her service in education, philanthropy, and leadership roles, including serving as a Life Trustee for the University of Chicago.

AWARDS AND RECOGNITION

Throughout Dr. Reatha Clark King's life and career, she received several awards and recognition for her trailblazing work in science, education, leadership, and community service. They include:

- Inducted into the National Women's Hall of Fame
- Minnesota's Top 100 Most Influential People
- University of Chicago Alumni Award for Excellence
- General Mills Community Leadership Award
- Outstanding Woman of the Year Award
- Lifetime Achievement Award from the National Association of Black Chemists and Chemical Engineers
- The Charles E. Bowers Award for Leadership
- Minnesota's Black History Makers Award
- Leadership Award from the Women's Foundation of Minnesota
- Award of Excellence from the National Bureau of Standards

REFERENCES

- **Reatha Clark King: RAW in Her Own Words (Recorded 12-31-2022)**
King, S. C. (2022, December 31). *RAW in her own words* [Recorded interview].

- **Find a Trail or Blaze One: A Biography of Dr. Reatha Clark King**
Leibfried, K. (2022). *Find a trail or blaze one: A biography of Dr. Reatha Clark King*. Beaver's Pond Press. ISBN: 9781643437002.

- University of Chicago. (n.d.). *Trailblazers: Reatha Clark King*. Retrieved from https://www.trailblazers.psd.uchicago.edu/reatha-clark-king

- The HistoryMakers. (n.d.). *Biography: Reatha Clark King*. Retrieved from https://www.thehistorymakers.org/biography/reatha-clark-king-42

- Minnesota Science and Technology Hall of Fame. (n.d.). *Reatha Clark King*. Retrieved from https://www.msthalloffame.org/reatha_clark_king.htm

GLOSSARY

✦ **Sharecropper** – A farmer who rents land and pays the owner with a portion of the crops they grow instead of money. Many African American families in the South worked as sharecroppers after slavery ended.

✦ **Cotton Picking** – The process of gathering cotton by hand from the fields before machines were used. It was hard work, and many families, including Reatha's, picked cotton to make a living.

✦ **Segregation** – A system that kept Black and white people apart in schools, restaurants, and other public places. It was unfair and made life harder for African Americans.

✦ **Negro History Week** – A special week started in 1926 to celebrate the achievements of Black people. It later became Black History Month, which is now celebrated every February.

✦ **Perseverance** – Working hard and never giving up, even when things are difficult. Reatha showed perseverance in her education and career.

✦ **Rocket Fuel** – A special kind of fuel that helps rockets blast off into space. As a chemist, Reatha helped develop rocket fuels that were used in space exploration.

- **NASA (National Aeronautics and Space Administration)** – The U.S. government agency that explores space. Reatha worked as a chemist helping to develop fuel for rockets.

- **NBS/NIST (National Bureau of Standards/National Institute of Standards and Technology)** – A U.S. government agency that sets standards for science and technology. It was called NBS when Reatha worked there but is now known as NIST.

- **Philanthropy** – The act of giving time, money, or resources to help others. As a leader, Reatha made sure companies supported education and communities.

- **Trailblazer** – Someone who is the first to do something, paving the way for others. Reatha was a trailblazer as a scientist, university president, and corporate leader.

- **Valedictorian** – The student with the highest grades in a graduating class. Reatha's love for learning led her to become valedictorian of her high school.

- **HBCU (Historically Black Colleges and Universities)** – Schools started to give African Americans a chance to get a higher education when other schools wouldn't accept them.

- **Clark College** – A historically Black college in Atlanta, Georgia, where Reatha studied chemistry. It later merged with Atlanta University to form Clark Atlanta University in 1988.

- **Community Leader** – A person who helps improve their town or city by making positive changes. Reatha served on many boards and helped create opportunities for others.

- **MBA (Master of Business Administration)** – A graduate degree in business and leadership. Reatha earned an MBA to expand her skills in management and finance.

- **College Dean** – A leader in a university who oversees a department or group of programs. Reatha served as a college dean, helping students and faculty succeed in their education and research.

- **General Mills** – A company that makes popular foods like Cheerios, Pillsbury biscuits, and Häagen-Dazs ice cream. Reatha worked at General Mills, where she helped lead efforts to support education, health, and the arts.

- **General Mills Foundation** – The part of General Mills that donates money and resources to help communities. Reatha led this foundation, making sure schools, health programs, and arts organizations received support.

ACKNOWLEDGEMENTS

First, I thank God for His guidance and grace. To my family your love and support made this journey possible.

Deep gratitude to Dr. Reatha Clark King for her blessing to share her story and to Scott Clark King for his insights and the powerful *RAW in Her Own Words* interview.

A heartfelt thank you to Lion's Historian Press team, Nassim Malik Sarkar and Michael Nyame. Letitia deGraft Okyere for copyediting, and Kate Leibfried for the use of her invaluable research and editorial review.

To Amina Yaqoob, whose beautiful illustrations bring this story to life, thank you. And to everyone who supported this project, I am forever grateful.

ABOUT THE AUTHOR

Rosemond Sarpong Owens is the founder of Lion's Historian Press, a publishing house dedicated to amplifying voices that have been missing from literature for centuries. Inspired by the African proverb, *"Until the lions have their own historians, the tale of the hunt will always glorify the hunter,"* she is on a mission to change the narrative and create a more inclusive and diverse literary landscape. Rosemond, a mother of three girls who love to read, hopes this book will encourage children to embrace their heritage, be proud of who they are, and always dream big. Through her work, she strives to inspire young readers to see their unique stories reflected in the world and to believe in the power of their own dreams.

OTHER BOOKS IN THE HERITAGE COLLECTION

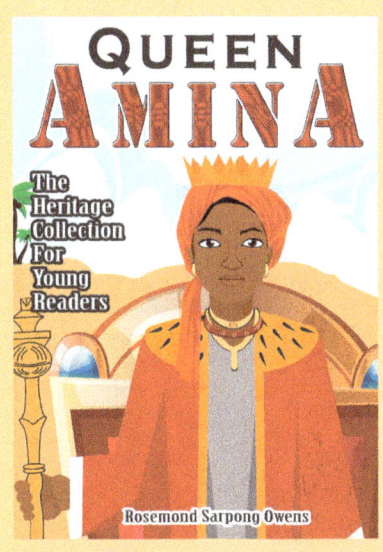
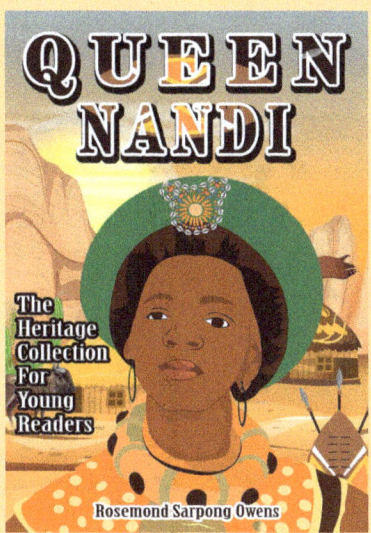
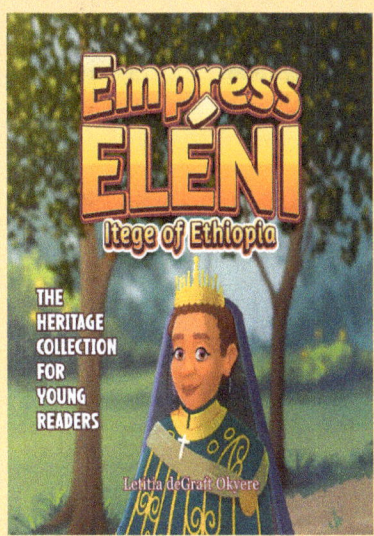

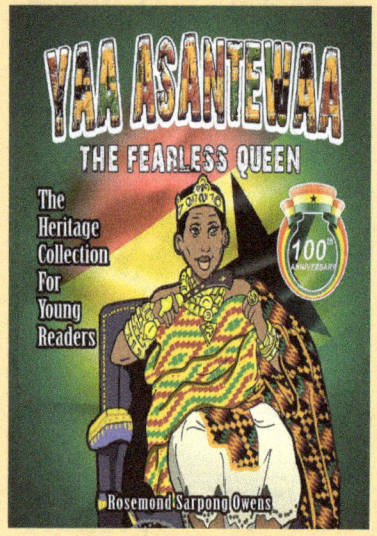
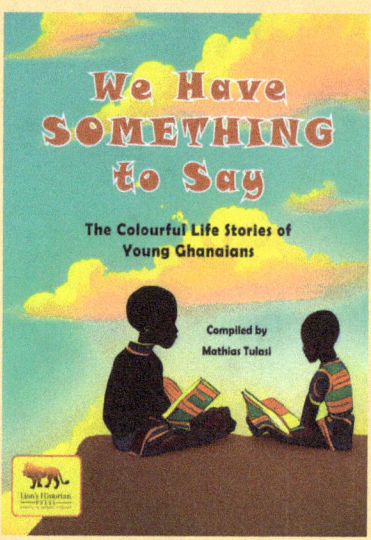
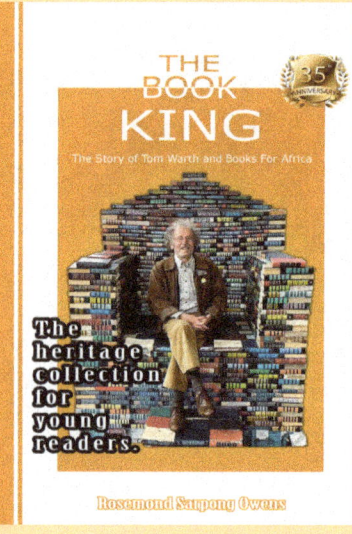

www.ingramcontent.com/pod-product-compliance
Lightning Source LLC
Chambersburg PA
CBHW060414010526
44107CB00006B/690